# Lippisch P13a Experimental DM-1

## Hans-Peter Dabrowski

**SCHIFFER MILITARY HISTORY**

Atglen, PA

**Sources:**
Lippisch — Ein Dreieck fliegt
Lippisch — Erinnerungen
Zacher — Studenten forschen, bauen, fliegen
Kens/Nowarra — Die detschen Flugzeutge 1933-1945
Masters — German Jet Genesis
Ford — Die deutschen Geheimwaffen
Peter — Der Flugzeugschlepp
Die Entwicklung des Deltaflügels/Aerokurier March 1968
Wenig bekannte deutsche Flugmuster/Aero February 1953
Aus den Geheimfächern der deutschen Luftfahrtindustrie/XXI. Folge — "Der Flieger"
Lippisch, Gluhareff, and Jones: The Emergence of the Delta Planform and the Origins of the Sweptwing in the United States/Aerospace Historian, March 1979
DM-1 Baubeschreibung, Berechnung und Betriebsanweisung (excerpts) 1945
NACA-RM No. L6K20, 1947
NACA-RM No. L7F16, 1947
Langley Air Force Base, Virginia, DM-1 Glider Disposal
History Development and Background (Lippisch to the Smithsonian Institution 1961)

Translated from the German by Don Cox.

Printed in the United States of America.
ISBN: 0-88740-479-0

This title was originally published under the title, *Uberschalljäger Lippisch P13a und Versuchsgleiter DM-1*, by Podzun-Pallas Verlag, Friedberg.

Published by Schiffer Publishing, Ltd.
77 Lower Valley Road
Atglen, PA 19310
Please write for a free catalog.
This book may be purchased from the publisher.
Please include $2.95 postage.
Try your bookstore first.

We are interested in hearing from authors
with book ideas on related subjects.

**Photos and documents were provided by:**
Herbert Dieks, Hans Justus Meier, Klaus Metzner, Hermann Nenninger, Heinz J. Nowarra, Reinhard Roeser, Peter F. Selinger, Günter Sengfelder, Fritz Trenkle, Hans Zacher as well as Smithsonian Institution, Washington, D.C., General Dynamics, San Diego, CA, and NASA Langley Research Center, Hampton Virginia. Thanks are expressed to all the above here.

The Me 163 rocket fighter — Lippisch's best known aircraft and the only aircraft of this type to enter series production. On October 2nd, 1941, Heini Dittmar reached a speed of 1003 km/h in this aircraft type.

# Development

Around 1930 Alexander M. Lippisch began working on the design of delta/tailless aircraft. Among others, there was the Delta I (1930), Delta II (1932), Delta III (1931/1932, built by Focke-Wulf in Bremen), Delta IV (1932, built by Fieseler as the F 3 "Wespe"), followed later by the Deutsche Forschungsanstalt für Segelflug (DFS) Delta IVc as the DFS 39 (1936) and Delta V as the DFS 40 (1937/1938). The Messerschmitt Me 163 "Komet" rocket-powered fighter stemmed directly from the DFS 39 and (other than his gliders) was the only aircraft by Lippisch produced in series — it was also the first and last rocket fighter ever mass-produced. This "firebird" proved to be a real, although rare, shock for the American bomber crews: something this fast, guided by the hand of man through the air had not existed up to this time. Me 163 test pilot Heini Dittmar reached a speed of over 1000 km/h in this aircraft on 2 October 1941 and became the first man to attain such a high speed. Because of the war, however, this feat was kept secret.

From 1943 onward Lippisch, as chief of the Luftfahrtforschungsanstalt Wien (LFW), continued to be involved with the problems of tailless aircraft in the supersonic range. His work aroused considerable interest in the Reichsluftfahrtministerium (RLM).

The last desperate attempts by the Luftwaffe to stop the Allied bomber streams were exemplified by the demand: fighters, fighters, fighters — quick and simple to build, cheap and from easily accessible materials, small dimensions, superior speed compared to enemy escort fighters and firepower. Nearly all aircraft manufacturers and designers in the Third Reich put designs to paper along

Fieseler Fi 103, known as the V1, was equipped with a Schmidt-Argus pulse engine and took off with the aid of a catapult which gave the flying bomb the necessary initial acceleration.

these lines. Even Dr. Lippisch's ideas during the last year of the war fit into this concept. Of his numerous variants, the Project 13a was actually given serious consideration.

Beginning with Project 12, his thoughts took concrete form with regards to the principle of using the ramjet for supersonic flight. In this case, the entire aircraft basically consisted solely of engine and fuel tank with some room provided for the pilot. The pulse-engine from Schmidt-Argus, similar to the ramjet, was certainly not a new concept, and the Fieseler Fi 103 flying bomb — better known as the V1 — flew in massive numbers against England with this type of engine. This engine (called the "Lorin" engine after

its inventor) with the wings of the Li P12/13, which could be controlled by flaps in the exhaust flow and ultimately was a flying wing, was called a "Triebflügel" (powered wing) by Lippisch. This was not to be confused with the ramjet-powered (Lorin) "Triebflugel" project of Focke-Wulf! This was a separate helicopter-like vertical takeoff aircraft with everything else operating as with a flying wing.

The ramjet operated by itself only after a certain speed had been reached. This process involved ram air entering in through an opening in the front, being mixed with fuel, compressed, ignited and forced out the back. The V1 in flight sounded somewhat like a

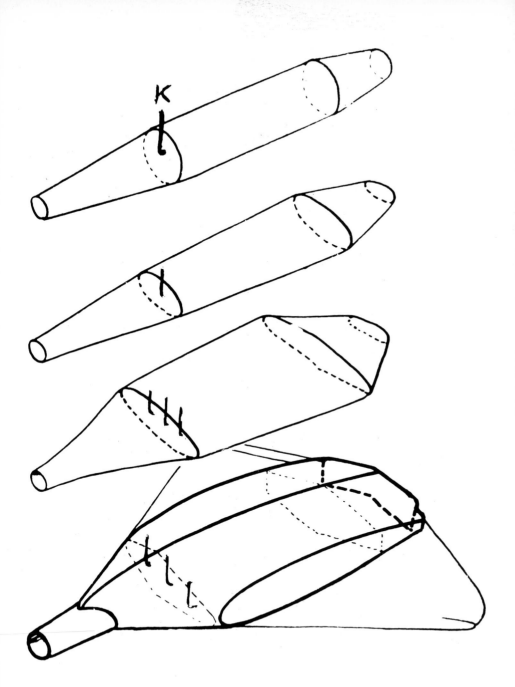

The stages from ramjet engine to powered wing, according to
Dr. Lippisch (K = fuel injection)

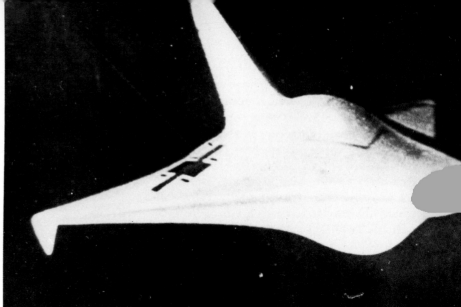

The Lippisch P12, powered wing design and predecessor to the
P13 (1943/1944)

tractor. There were no complicated, moving parts and such a thruster was quick and cheap to build. The aircraft being proposed by Lippisch would be powered in part by coal as fuel — a "flying coal oven" which would also be able to reach supersonic speeds.

Refined coal and heavy oil injected in was to be burned: for a flight of 30 minutes approximately 500 kg of coal granules was necessary. Tests in this area were conducted by Dr. Schwabl (Vienna) and Dr Sänger (DFS).

Briefly described, the Lippisch Project 13a was a supersonic fighter of metal construction. cantilever delta wing with a 60 degree swept leading edge, profile thickness of 15 percent, ramjet engine installed in center portion of wing. Wingspan was 6.00 meters, length 6.70 meters, height 3.25 meters. The control rudder was a triangular fin fixed into the center of the wing, also with a 60 swept leading edge, profile thickness 17.5 percent, wing and rudder leading edges were rounded off, wing surface was 20.0 m2. No landing gear, but a central skid for landing. As armament, the MK 108 (caliber 30 mm) then common to all jet and rocket fighters would have been utilized.

Coal burning ramjet as engine and rocket motor for takeoff and for reaching the necessary minimum speed for the main engine. The pilot was to sit inside the vertical stabilizer, which had a part of its leading edge glazed.

In order to study the flight characteristics of this unusual design, free flights were conducted using a scaled-down version of the P13a beginning in May 1944 at Spitzerberg near Vienna. A wind tunnel model as a pure triangular design was also built (August 1944), which was studied in the supersonic wind tunnel of the Aerodynamischen Ver-

The Focke-Wulf powered wing design, powered by thre Lorin ramjet engines. The required initial velocity was to be provided by rockets mounted in the engines.

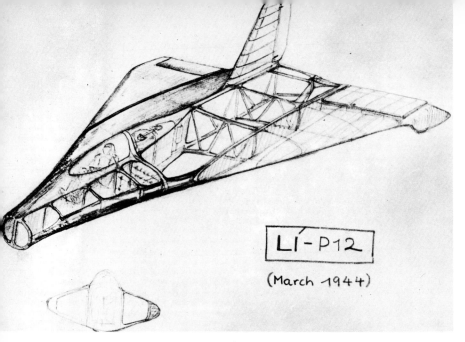

LÍ-P12

(March 1944)

A variant of the Li P12 project.

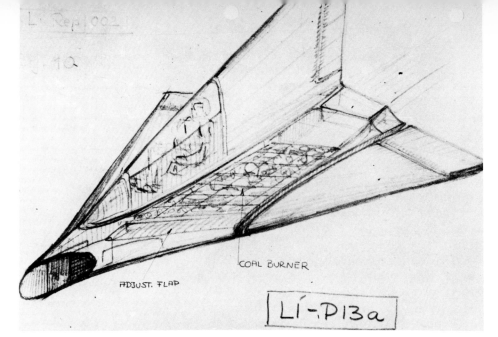

ADJUST. FLAP

COAL BURNER

LÍ-P13a

One of several drafts of the Li P13a project.

With Lippisch the design for a project was usually altered many times — often there were considerable differences to the basic design.

A cutaway of the final shape of the P13a, which is nearly identical to the DM-1 glider.

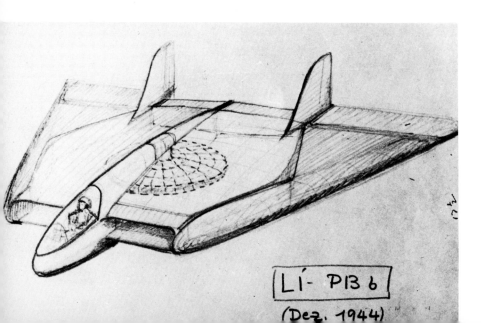

LÍ- P13b

(Dez. 1944)

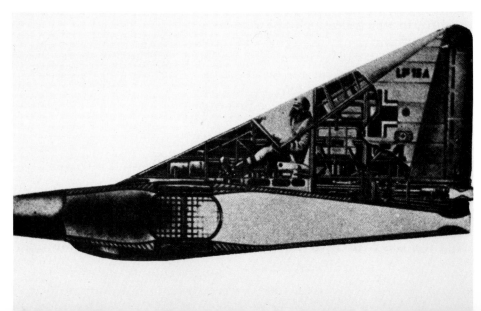

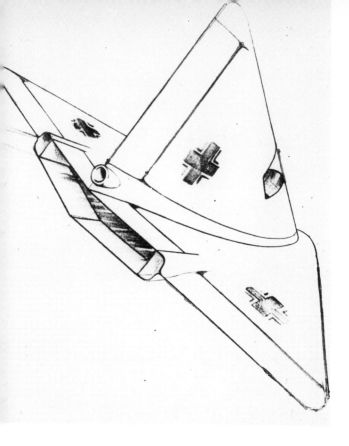

This sketch shows the position of the rocket engine and the wide exhaust opening, in which steering flaps were to be operated.

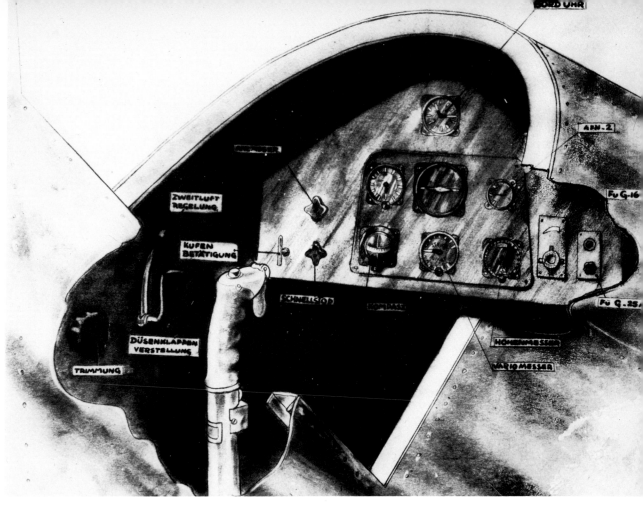

The inner cockpit of the Lippisch Project P13a.

suchsanstalt (AVA) in Göttingen. After a successful conclusion of these studies more exacting research of the P13a's flight characteristics in its original size was to be done using a non-engined, but flyable and manned 1:1 model built of wood. And this test glider was later called the DM-1.

Due to the assistance Wolfgang Heinemann, an AKA pilot from Darmstadt, had given to Lippisch in the LFW the Flugtechnische Fachgruppe (FFG) Darmstadt was given the contract to build the 1:1 scale model. Such contracts were very important for holding the academic flying groups

together, as far as this was possible during the war.

Thus, beginning in August/September 1944, work was begun there in conjunction with Lippisch and according to his proposals. This was after three versions had been more closely examined, the one with the large

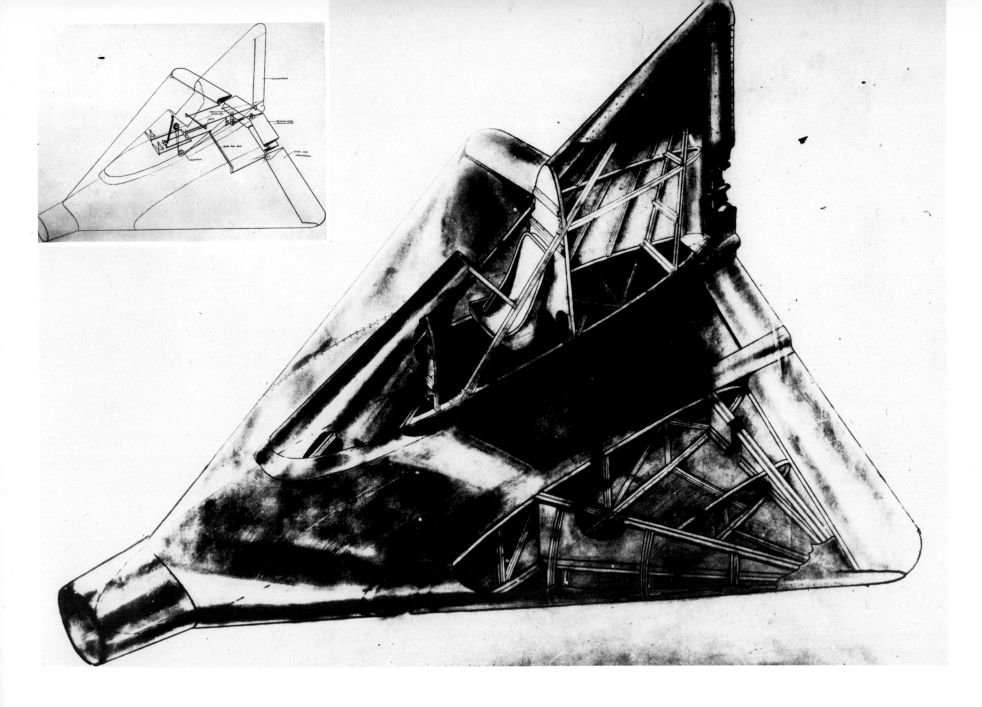

Construction and control surfaces of the Li P13a.

vertical stabilizer eventually being selected. The aircraft was given the designation D33; when asked about this subject after the war, Lippisch was of the opinion that the correct designation should have been Lippisch P13a V-1 (as the predecessor of the planned P13a fighter).

On 11/12 September 1944 the FFG Darmstadt was destroyed in a bombing attack and everything pertaining to aircraft and materials which could be saved was sent elsewhere. According to Leo Schmidt (of the DVL, responsible for the management of the aeronautical studies groups such as the academic flying groups, or "Akafliegs", as they were known in the Third Reich), the D-33 then under construction was transported to Prien on the Chiemsee and work was continued at the FFG München facilities. The "spiritual father" of the test glider no longer concerned himself with the design progress. The aircraft was now called the DM-1 (D = Darmstadt, M = Munich).

Since the aircraft was still without any type of internal power, it was to be carried piggyback, anchored at three points on a twin-engined Siebel Si 204 A to altitude and there, released along its flight path, would reach high speeds. With the aid of additional solid propellant rockets mounted on the Mistel fittings in front of the main wheels it was hoped to reach speeds of 800 km/h. In 1945 two brand new Siebel 204 A planes stood at the edge of the airfield in Prien, ready for use.

Hans Zacher, who shortly before the end of the war came from Ainring (DFS), was responsible for the flight mechanics/flight handling and the planned piggyback flight, and possibly would function as the pilot of the test glider. In any case, there was no flight testing actually ever conducted.

Briefly described, the DM-1 was a single

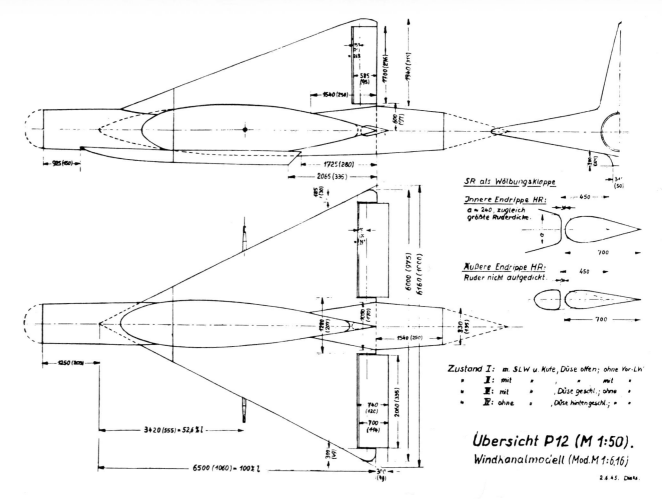

Above: Drawing of the wind tunnel model, here designated as the P12, probably because it was built before the P13.

The wind tunnel model actually built had stabilizing fins on the nose and was covered with woolen fabric strips for monitoring air flow.

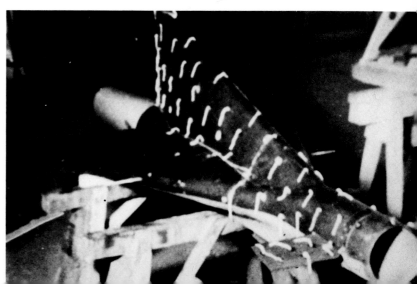

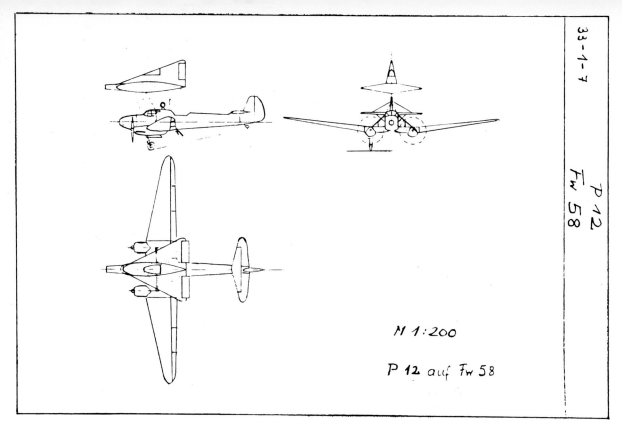

P 12
Fw 58

M 1:200

P 12 auf Fw 58

Another P12 1:1 scale mockup was to be tested practically in a Mistel arrangement on an Fw 58.

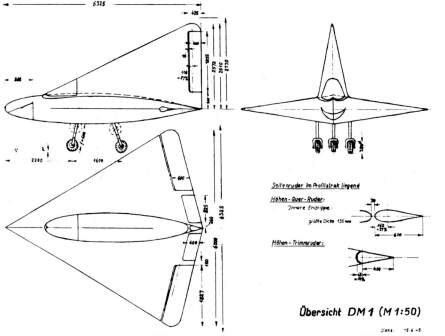

*Seitenruder im Profilstrak liegend*

*Höhen-Quer-Ruder:*
*Innere Endrippe:*
*größte Dicke 130 mm*

*Höhen-Trimmruder:*

**Übersicht DM 1 (M 1:50)**

The original form of the DM-1. This drawing was completed for the Americans on 15 June 1945.

seat delta test aircraft (for researching the flight characteristics of the P13a supersonic fighter) made of wood, plywood and steel tubing. One-piece wing of open rib design, cantilever. Elliptical-symmetrical profile by Dr. Ringleb, 15 percent thickness, vertical and horizontal stabilizer surfaces were hinged, with trim tabs in the interior portion. There was no fuselage in the normal sense, the cockpit was located in part within the forward portion of the vertical stabilizer, in part just behind the forward leading edge between the two main spars. A window in the lower forward floor provided better view at high angles of attack (during landing at roughly 35 degrees). The triangular vertical stabilizer had a similar profile as the wing. Trimming was accomplished by a manually-operated water transfer pump (35 liters) from a rear tank to a nose tank and back. The tricycle landing gear, which could only be retracted on the ground, had a 60 cm (!) travel stroke. Width was 6.00 meters, length 6.32 meters, height 3.18 meters, wing surface area 199. m2, wing loading 23 kg/m2, weight empty 375 kg, gross weight 460 kg, best glide ratio = 7, calculated maximum speed along the flight path 560 km/h, landing speed 72 km/h. No type of armament was planned for this test version; overall, the DM-1 was kept to simple, spartan standards.

The DM-1 under construction: connecting the main spar with the vertical stabilizer.

The vertical stabilizer under construction.

View into the interior: the ribs.

The vertical stabilizer: the design principle is recognizable here — radiating bowed strips tapering off towards their ends.

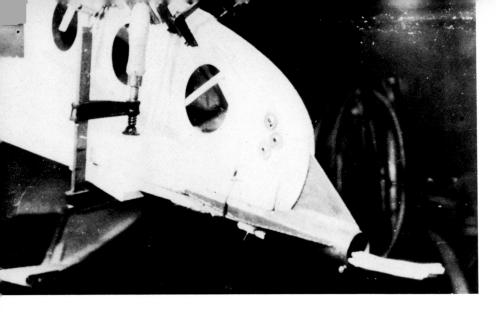

Forward fuselage: center rib with bottom of water tank.

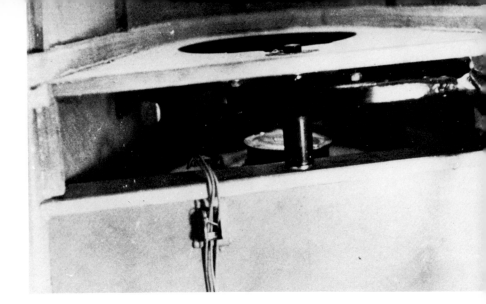

Rudder drive and water tank in the rear fin. The electrical connections lead to the tank gauge.

Installation of the landing gear between the main ribs. The drive for the elevator trim runs over the rollers (center of photo).

Planking the underside. At work are the cutters from FFG Aachen.

Adjusting the horizontal stabilizers.

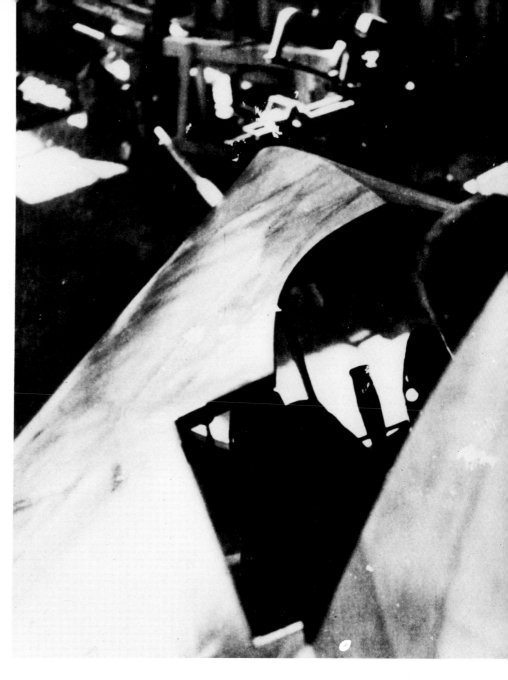

View from above into the cockpit with floor glazing.

Airframe minus fuselage installation, seen from the front and below.

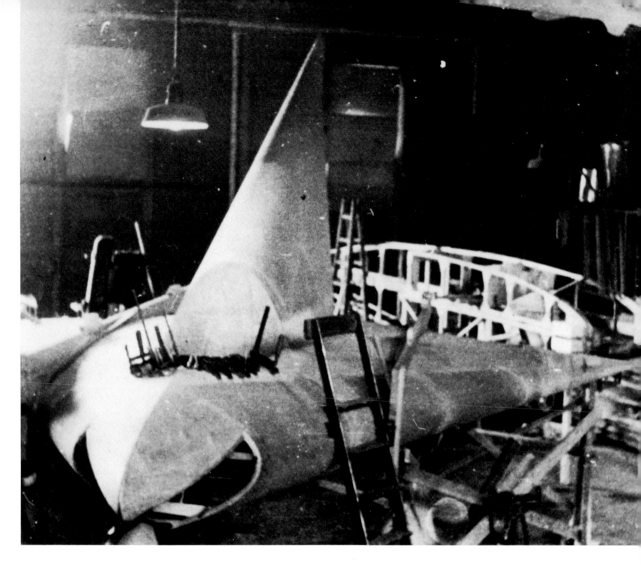

Above: Airframe of the DM-1 from above. In the background can be seen the mockup for the DM-2/DM-3 under construction.

Left: Canopy glass is being finished. At work is Paul Stöhr (right) and Hermann Nenninger (left).

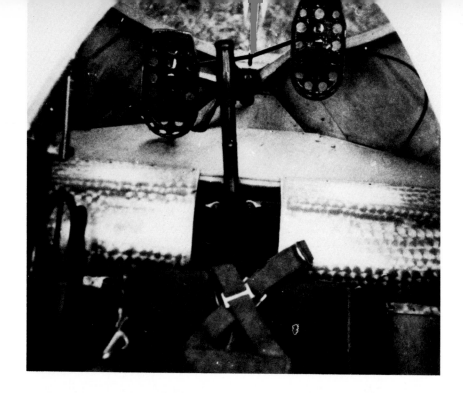

Nosewheel (the landing gear utilized standard 350x50 wheels), Mistel bolts, air intake.

View from above into the cockpit. The pedals and the pilot's seat stemmed from a Bücker Bü 181.

Left: Hans Zacher in the DM-1; the canopy opened to the right (as seen from the direction of flight) and in case of emergency could be removed entirely.

View from below through the floor glazing into the cockpit.

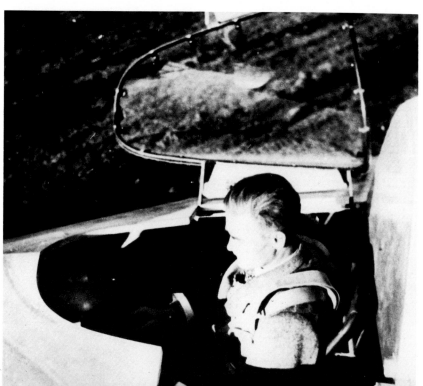

# The War Is Over —
# The DM-1 Continues
# To Be Built!

The war was almost over as the Prien am Chiemsee airfield was occupied by American troops on 3 May 1945. There they discovered the DM-1 under construction. From May 1945 this unique bird was strictly guarded and it was planned to continue construction. On 9 May 1945 there was a visit by the commanding general of the US 7th Army, Patton, with a large entourage. Continued construction was arranged. The project officer was a certain Major A.C. Halzen from the Air Technical Intelligence Section USAFE, a likeable man with whom it was easy to deal. Professor Theodore von Kármán, under the auspices of the USAF Scientific Advisory Groun (SAG), pressed decisively for the continued development of the DM-1.

Thus it was that building continued throughout the summer of 1945, with many people desiring to see the secret bird during this period. Even the famous ocean pilot Charles Lindbergh came. He had established friendly relations during the mid-1930s in Prien and he therefore became the first who broke through the order "No Fraternization", stretching his hand out to the Germans.

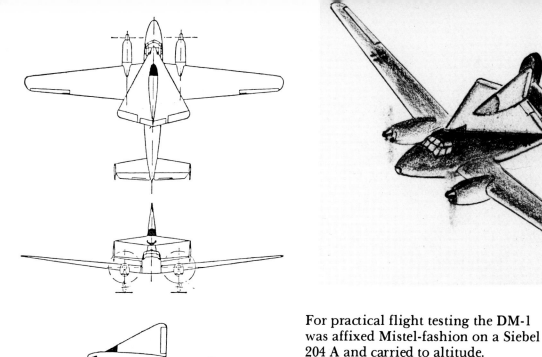

For practical flight testing the DM-1 was affixed Mistel-fashion on a Siebel 204 A and carried to altitude.

Right: Dr. Lippisch's assistant Wolfgang Heinemann next to the DM-1.

Here it's clearly noticeable, how spartan and simple the DM-1 was designed: from the cockpit a person could see directly into the interior of the wing.

The DM-1 after completion, at the Prien airfield.

Sitting in front of the American high-wing aircraft in the background, it's clear how much the shape of the DM-1 was ahead of its time.

The DM-1 seen from the front.

Rudder and horizontal stabilizer seen from below.

Right: In front of the Prien FFG hangar in Munich.

Remembrance photo: left to right — Herbert Diecks (DA), Wolfgang Heinemann (DA), Klaus Metzner (MU), Hermann Nenninger (MU). Not seen in this photo: Bernd Kulartz (MU), Hans Zacher (DA) and Alfred Henschel of the Flugtechnische Arbeitsgemeinschaft Chemnitz.

Prien airfield: in the background is the Douglas DC 3 (C-47) planned for Mistel towing.

Above and below: The DM-1 set on its rear; this is roughly the same angle-of-attack as during landing.

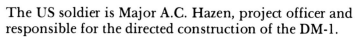

The US soldier is Major A.C. Hazen, project officer and responsible for the directed construction of the DM-1.

The DM-1 was only coated with a colorless weather protective paint; additionally there was no filler material used in order to cover and smooth over the nail holes in the planking.

# The DM-1 Travels To America

Under the supervision of the Americans the DM-1 was to be tested practically in Germany. Of course, there weren't any more Siebels, but a Douglas C 47 (DC-3) was to carry the wooden flyer piggyback to altitude. A corresponding plane was already planned for this purpose. But this was not to be: a quick decision was made that the DM-1 was to be thoroughly examined in the USA. A wooden crate was specially made "dimensioned" (6.5 x 6.5 x 3.0 meters), in which the DM-1 could be stored in one piece. The wing tips were to be removed for crating and the landing gear retracted. Finally, the craft was wrapped in a weatherproof foil and "vacuum sealed" with the aid of a modified vacuum cleaner. Together with six hundredweights of salt (for removing moisture during the planned ocean voyage), the glider disappeared into the crate. The Americans even "paid" for the aircraft by a corresponding voucher to the German authorities (credit toward reparation contributions).

On November 9, 1945, the designated crate was loaded onto a lowboy trailer of the US Army and began the several thousand kilometer journey to America. Nevertheless, immense problems sprang up beginning with the first kilometer.

Despite having previously measured all obstacles along the route, the journey got hung up at a railroad underpass. It was necessary to unload the crate from the trailer and pull it through the underpass millimeter by millimeter using a crane truck. By autobahn it was transported to Mannheim and then down the Rhine to Rotterdam, con-

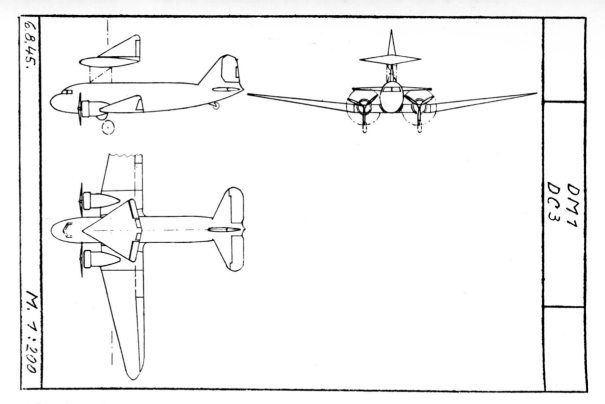

The planned Mistel flight did not take place on the part of the Americans, either. Due to the Allied Control Council's determination German pilots were restricted from flying, and he wooden bird was probably rather mysterious for the Americans, so that it was never actually flown (as shown in this photo montage).

Construction of the "dimensionally cut" transport crate. The DM-1 was fitted in (the rudder end piece is removed) and finally covered with a weatherproof foil; the air is then removed with the aid of a vacuum cleaner (below right)

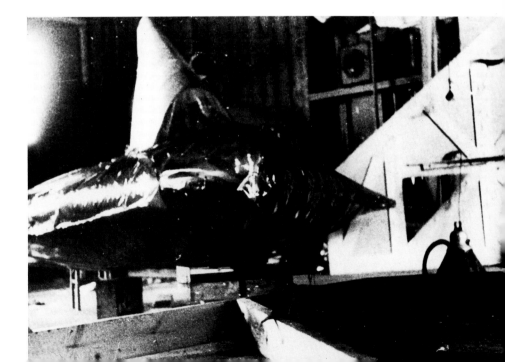

Model of the transport crate, seen from above.

Now the flyer disappears into the crate, which is then lifted out of the hangar by a US Army crane truck.

tinuing onward on board the SS "King Hathaway" to Boston (arriving 19 January 1946). From there, the coal freighter "Bomair Seam" took it to Norfolk. The end point was the "Full Scale Wind Tunnel" at Langley Field in Virginia. As early as two days after the arrival in Boston the Air Force Material Command contracted with NACA (NACA = National Advisory for Aeronautics, precursor to today's NASA) to test the DM-1 in this huge wind tunnel. A test program, which was to evaluate all possible flight characteristics of this test glider, was worked out between this Air Material Command and Professor Kármán. Two NACA reports, RM No. L6K20 and L7F16 thoroughly document

The transport crate is carefully loaded onto a lowboy trailer. The English writing on the crate clearly emphasizes its extremely sensitive contents.

The road led through Prien. On both sides of the crate a close likeness of Wilhelm Busch's bad luck raven "Huckebein" was painted, the group's mascot and symbol of the Munich Aka pilots.

The journey came to a halt after just a few hundred meters: the lowboy was not able to fit beneath the Seestrasse railroad underpass. Captain Landis, who was responsible for the transport and had measured all obstacles in advance, could not comprehend the problem and "broke down" in despair over the crate.

The crate was taken off the lowboy. There developed a feverish pace, which was observed by interested citizens of Prien. Here and there could be heard the clicks of cameras, but the participants probably didn't have the most friendly comments regarding this misfortune.

The transport crate was said to have been painted in gray, writing in white and the sign "USE NO HOOKS" was in red. One didn't want to unnecessarily give the impression that this was a military object.

Millimeter by millimeter — then success and the journey continues.

Last view of the disappearing crate, autobahn stretch Prien/Bernau, westward.

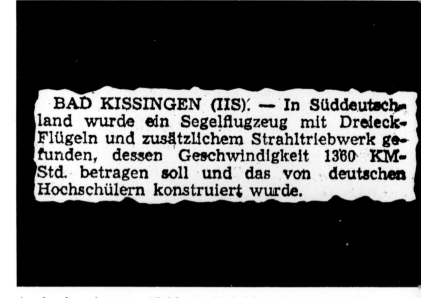

BAD KISSINGEN (IIS). — In Süddeutschland wurde ein Segelflugzeug mit Dreieck-Flügeln und zusätzlichem Strahltriebwerk gefunden, dessen Geschwindigkeit 1360 KM-Std. betragen soll und das von deutschen Hochschülern konstruiert wurde.

A rather deceptive report: "Salzburger Nachrichten" newspaper article from 5 November 1945. It reads: Bad Kissingen (IIS). A delta-wing glider aircraft and additional jet engine was found in southern Germany, said to be able to reach speeds of 1360 km/h and constructed by German graduate students.

The test glider after arrival in the Langley Memorial Aeronautical Laboratory, Langley Field, Virginia.

the results and these were, bluntly put, disappointing: lift and flow characteristics were very poor. Dr. Lippisch reported that he was approached and requested to explore the problems and provide assistance. He attributed the poor data to the influence of the so-called "Reynolds coefficient" (too early separation of the airflow). The anticipated improved results were achieved by fixing blade-like, sharp metallic strips onto the wing leading edges (for approximately half the total length of the wing).

By the way, neither NACA report cited the name of Lippisch anywhere. No German name was mentioned at all. The DM-1 is only briefly mentioned as a German development. During the course of wind tunnel testing the DM-1 was modified several times: the large vertical stabilizer was removed, the gaps between the vertical stabilizer and wing were glued over, the sharp edges were removed and reinstalled, etc. Finally, a smaller vertical stabilizer was installed along with a separate pilot's canopy; at the same time the aircraft was tapered more to the front with corresponding attachments. These versions provided the better results.

The **DM-1** with its 6 meter wingspan looks like a toy inside the giant wind tunnel.

NATIONAL ADVISORY COMMITTEE FOR AERONAUTICS

RESEARCH AUTHORIZATION     No. 1406

Full-Scale Wind Tunnel Investi-    By   Langley Memorial Aeronautical
tion of Lippisch DM-1 Supersonic       Laboratory.
...der.

Approved ................... , 194...

                Chairman, Subcommittee on ...................

Issued
Approved   February 5 , 1946      J. W. CROWLEY
                             Chairman, Executive Committee
            Acting Director of Aeronautical Research.

In accordance with authority of Executive Committee March 19, 1942.

...ose of investigation (Why?)

     To determine the low-speed stability and control character-
istics of the Lippisch DM-1 supersonic glider.

...description of method (How?)

     The Lippisch DM-1 supersonic glider will be mounted in the
test section of the Langley full-scale wind tunnel and, at a
normal C. G., in the flight condition   Information will be
obtained, as follows: A. Lift coefficient, drag coefficient and
pitching moment coefficient versus angle of attack with elevator
free and at a number of elevator deflections throughout the elevator
deflection range; B. Rolling moment, yawing moment and pitching
moment versus angle-of-yaw, with rudder free and at various rudder
deflections, at several angles of attack covering the speed range;
C. Rolling moment versus aileron deflection; D. Control forces
arising in the preceding tests.

...arks:

     Requested by the AAFATSC in letr. dated 21 January 1946,
TSEMO(TSMTF-2)FHH/bm

...s of reports ................      Publications ................

AS 320-1 (1406)

                        Completed ................... , 194...

... No. 18        U S GOVERNMENT PRINTING OFFICE

In the wind tunnel.

Testing the landing approach angle.

The contract for testing the DM-1 at Langley Field.

The test glider covered with woolen fabric strips.

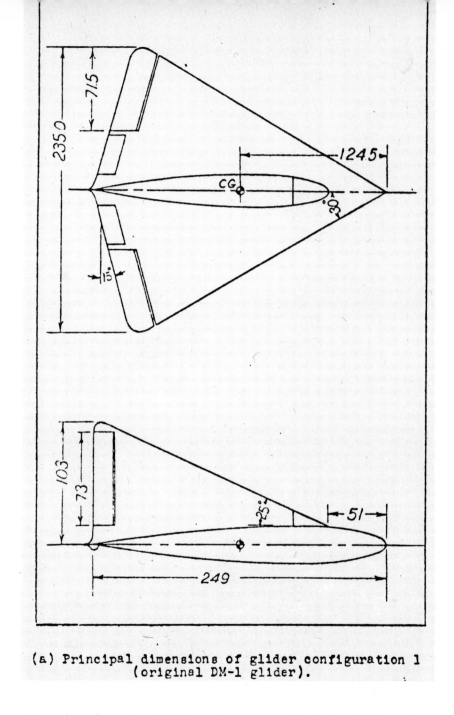

Section AA

Section BB

(a) Principal dimensions of glider configuration 1 (original DM-1 glider).

(b) Dimensions of the semispan sharp leading edges and of the elevon control-balance slots.

Drawings from the NACA study reports: left is the DM-1 in its original form, right with the "sharp edge" — measurements in inches.

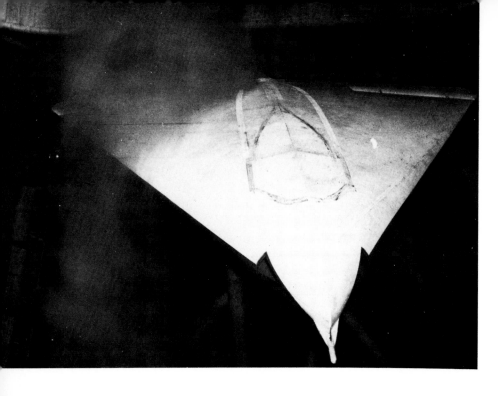

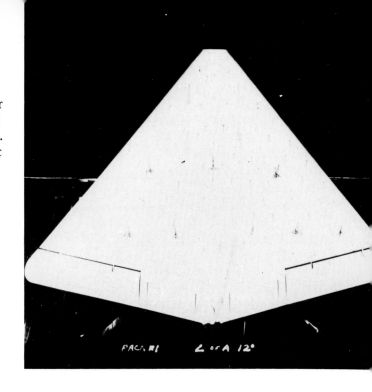

Above left and below: Smoke testing with rudder removed, with and minus sharp edges. above right: Fabric strip testing.

Right: Side view with sharp leading edges and missing rudder. Other than the wind tunnel testing, no other testing is known to have been done.

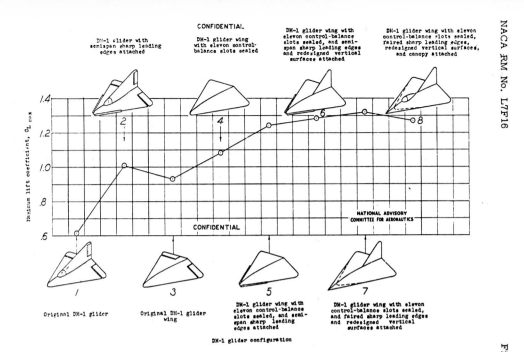

DM-1 glider with semispan sharp leading edges attached

DM-1 glider wing with elevon control-balance slots sealed

DM-1 glider wing with elevon control-balance slots sealed, and semispan sharp leading edges and redesigned vertical surfaces attached

DM-1 glider wing with elevon control-balance slots sealed, faired sharp leading edges, redesigned vertical surfaces, and canopy attached

Maximum lift coefficient, $c_{L_{max}}$

Original DM-1 glider

Original DM-1 glider wing

DM-1 glider wing with elevon control-balance slots sealed, and semispan sharp leading edges attached

DM-1 glider wing with elevon control-balance slots sealed, and faired sharp leading edges and redesigned vertical surfaces attached

DM-1 glider configuration

Figure 4.- Summary of the effects of the modifications made to the DM-1 glider on the maximum lift coefficient.

Fig. 4

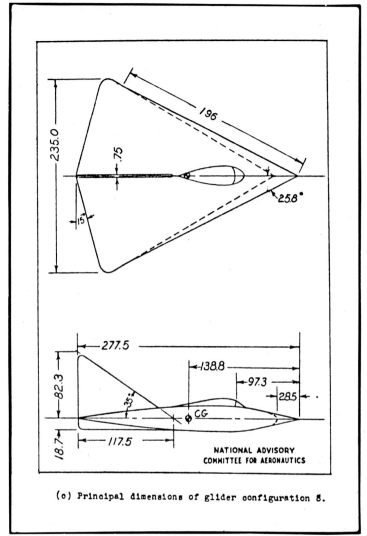

(c) Principal dimensions of glider configuration 8.

Above: The most favorable form for aircraft and pilot.

Above left: All tested configurations of the DM-1.

Left: Configuration #8 in the wind tunnel.

The DM-1 can hardly be recognized after modification.

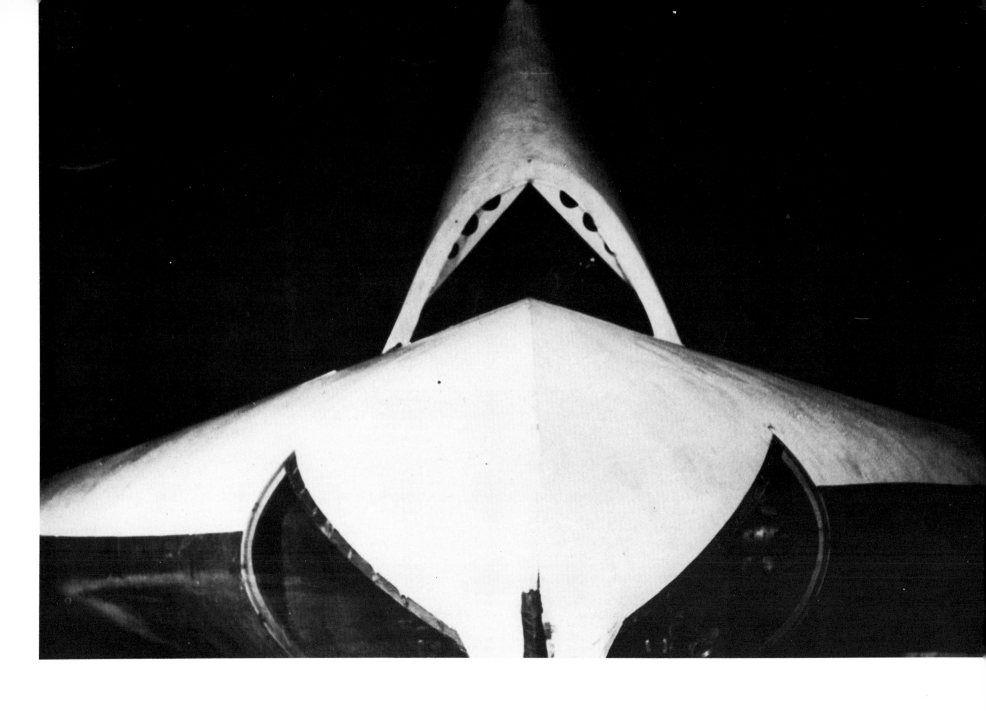

End of testing — the wooden bird has survived, albeit a bit worse for the wear.

# The Way To The Museum

The AMC NACA engineering field officer from Langley Field, Major Howard Goodell, who paved the way for the transfer to the National Aeronautical Museum in Washington, D.C., expressly pointed out that the DM-1 was not the first delta wing aircraft in the USA, as the Ludington-Griswold Company, Saybrook, Connecticut, had demonstrated a delta planform of Michael Gluhareff, a Russian exile living in America. This was only a wind tunnel model, so that it can be said that the DM-1 was the first true aircraft of this type — a milestone in wood.

From January 1948 the DM-1 had rested in a NACA warehouse at Langley Field. Apparently, the National Aeronautical Museum of the Smithsonian Institution was not interested in the worn out flyer. For NACA's part, it didn't consider the DM-1 as anything special, certainly not worthy of standing in what is probably the world's largest and most famous aviation museum.

There were suggestions on the least expensive way to transport it. That was in November 1949. In January 1950 the museum then received the DM-1. Nearly 20 years later Dr. Lippisch complained bitterly that, "it's still today sitting in a warehouse and rotting." Now, however, it is actually in the possession of the museum and awaits better days at the hands of the Paul E. Garber Preservation, Restoration, and Storage Facility in Maryland. Whether and when this predecessor of all modern delta-winged aircraft will be restored and made available for public viewing is written in the stars.

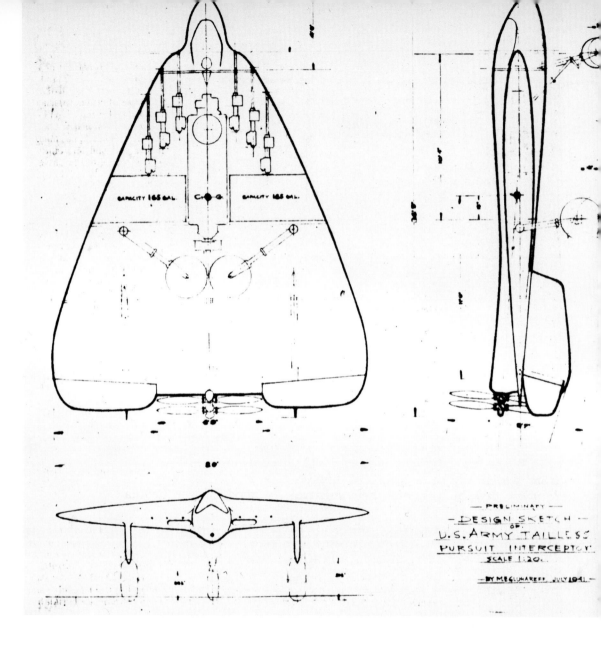

American delta-planform design by Michael Gluhareff.

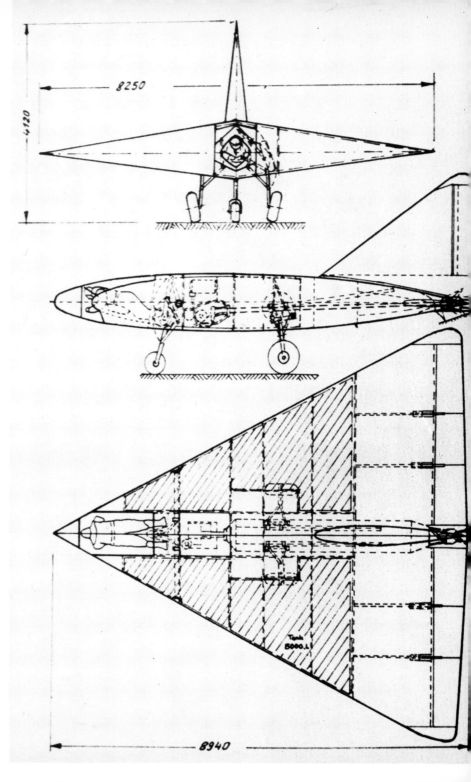

There were also follow-on developments (from which Lippisch distanced himself, claiming that he was not their spiritual father), also made of wood: the DM-2 with Walter rocket engine, glazed nose, pilot in prone position, overall larger dimensions with a smaller vertical stabilizer; the DM-3 as the DM-2, but with pressurized cockpit (the mockup was already under construction in Prien, but was not given further attention by the Americans); the DM-4 airframe as an engine testbed. The plans for these aircraft were stolen in Rosenheim from the project officer, Major Hazen. Shortly afterward, a Russian radio station was said to have claimed that the plans were now in Soviet hands.

Above and right: DM-2/DM-3 as design and model. Since the Aka pilots had carefully made copies of the plans, they were able to help Major Hazen out of his predicament.

Left: an entire DM "fleet" in model form, anno 1945. The large difference between the DM-2 and DM-1 is apparent.

Below: Major Hazen (at the wheel) enjoyed driving this mignonette green BMW requisitioned from a Bad Tölz dentist. This is most likely the car from which the plans were stolen.

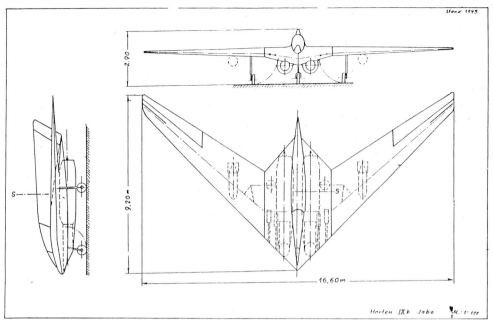

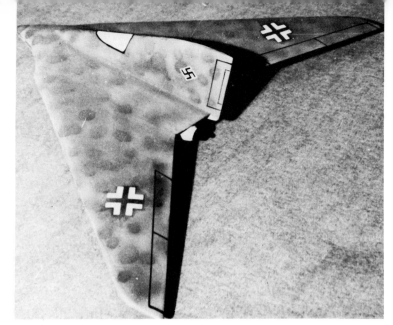

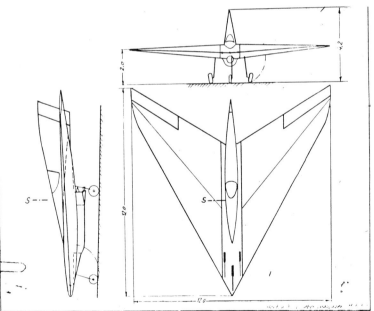

Aside from Alexander Lippisch the brothers Walter and Reimar Horten also pursued the idea of a flying wing. With their Horten IX B and Horten XIII jet fighter designs they came to the conclusion of stationing the pilot in the vertical stabilizer. Therefore, these birds bore some resemblance to the DM-1, although they did not progress beyond the project stage.

Above: Ho IX B, below Ho XIII B. Even here there were numerous variations and it's not certain that all were true Horten designs.

Dr. Alexander Lippisch, the "spiritual father" (2 November 1894 — 11 February 1976)

He began with sailplane building in 1922, from 1930 the first tailless aircraft, developed the Me 163 rocket fighter at Messerschmitt. At the end of the war, Lippisch fled from Vienna, which was being occupied by Soviet troops, to Strobl/Wolfgangsee and there fell into American hands. As early as the 23rd of May 1945 he was flown to Paris; there he gave a lecture before American specialists, but encountered little understanding. His emigration to the USA, planned for the fall of 1945, was initially postponed due to protests against the work of German scientists in the

Above: Dr. Lippisch with his designs in the USA (LIFE report on "Nazi Scientists")
Right: Prof. v. Kármán in Germany shortly following the capitulation.

States. So it was first off to Wimbledon (England).

In January 1946, meaning at almost the same time as the DM-1, Lippisch did indeed come to the USA, arriving at Wright Field in Dayton, Ohio.

The Convair aircraft company put Lippisch's delta idea into reality with the XF-92 — the first delta fighter flew!

**Professor Theodore v. Kármán**
**"Savior" of the DM-1**
**(11 May 1881 —7 May 1963)**
stemmed from Hungarian-Jewish ancestry, taught at the Technische Hochschule Aachen (expert on matters of airflow). As early as 1929/1930 he came to the United States as he noticed the climate in Germany growing worse for Jews. Taught at the Technical Institute in Pasadena, California and became a US citizen in 1935. Shortly before the end of the war he was in Paris and, immediately following the capitulation was in Germany in order to determine the state of German aeronautical science on behalf of the USAF. He was thus able to "interrogate" in Göttingen his old friend and teacher Ludwig Prandtl.

**AFTERWORD**

The Prien airfield no longer exists. Flight operations were brought to a halt in 1964. Delta fighters and bombers flew and continue to fly throughout the world, and even the American space program utilizes this design — but in the meantime the place where the "cradle" of this development rested has been converted to farmland.

## Technical Data (Estimated)

| | P13a | DM-1 | DM-2 | DM-3 | DM-4 |
|---|---|---|---|---|---|
| Width m | 6.00 | 6.00 | 8.25 | 8.25 | |
| Length m | 6.70 | 6.32 | 8.94 | 8.94 | |
| Height m | 3.25 | 3.25 | 4.12 | 4.12 | |
| Crew | 1 pilot sitting | 1 pilot sitting | 1 pilot prone | 1 pilot prone | |
| Function | Supersonic fighter | Test model | Supersonic test Model | DM-2 with pressurized cabin | Engine testbed |
| Climb rate* km/h | | – | 500 | | |
| Maximum speed* km/h | 1600 | 560 | 6000 | 10000 | |
| Landing speed* km/h | | 72 | 85 | | |
| Engine | Ramjet | – | Walter rocket | Walter C-rocket | Walter C-rocket |
| Empty weight kg | | 375 | | | 2500 without engine |
| Flying weight kg | 2300* | 460 | 11500* | | |
| Material composition | metal | wood steel tubing | wood, metal | wood, metal | |
| Planned armament | 2 x MK 108 | | | | |

*Calculated data

The Convair XF-92 A
delta-winged test
aircraft was a result of
the Lippisch concept of
a 60-degree delta wing
for high speeds.
Experiments with this
wing began at the
beginning of July
1946(!) at Convair, after
in-depth discussions
had been conducted
with Lippisch. The Li
P13 and the wind
tunnel testing with the
DM-1 played a
significant role here.
On 18 September 1948
the XF-92 A flew for the
first time as the first
delta jet airplane.

LIPPISCH DM-1

ISBN: 0-88740-479-0